Visual Category Theory

Dmitry Vostokov

Brick by Brick

$CoPart_1 \in C^{op}$

Visual Category Theory, CoPart 1: A Dual to Brick by Brick, Part 1

Published by OpenTask, Republic of Ireland

Copyright © 2021 by OpenTask

Copyright © 2021 by Dmitry Vostokov

All rights reserved. No part of this book may be reproduced, stored in a retrieval system, or transmitted, in any form or by any means, without the prior written permission of the publisher.

Product and company names mentioned in this book may be trademarks of their owners.

OpenTask books and magazines are available through booksellers and distributors worldwide. For further information or comments, send requests to press@opentask.com.

A CIP catalog record for this book is available from the British Library.

ISBN-13: 978-1912636815 (Paperback)

Revision 1.01 (February 2021)

Preface

After publishing some parts of the Visual Category Theory Brick by Brick series (there are 7 of them at the time of this writing with more than 110 pages with brick diagrams), I got suggestions from readers to include standard category language diagrams. Because the original series translated abstract categorical concepts into the language of LEGO® bricks, I realized that the opposite way of translating brick constructions to the standard diagram language of category theory would benefit comprehension of definitions.

Since usual categorical diagrams are black and white and occupy less space on paper, these CoParts also decided to include additional color-enhanced diagrams in the spirit of brick constructions when arrow source and target parts use different colors.

The series itself became a reference dictionary (at least for the author), with more than 70 definitions, so I believe these dual parts make perfect sense for learning. The main reason I created this series was to teach myself category theory beyond superficial textbook reading.

In these CoParts from CoSeries (named after opposite categories with reversed arrows), I keep the same 1-to-1 page correspondence between Parts and CoParts. Page layout is similar: location of explanatory notes (written using standard mathematical notation) is the same — only bricks are replaced by letters, dots, and arrows. Color-enhanced diagrams may also use additional symbols. Therefore, this CoSeries can be used independently from the original series.

A category *C* consists of two collections

X Y Z

any unnamed object: • $Ob(C)$ - a collection of objects

X Y Z

unnamed distinct objects:

f

g

any unnamed arrow:

⟶

unnamed distinct arrows:

Ar(*C*) - a collection of arrows
(also called morphisms)

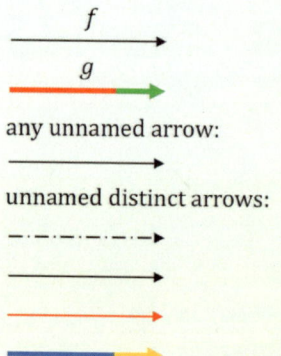

A   *B*   *D*   *E*

Any $X, Y \in Ob(\mathbf{C})$ have a set of arrows (maps, morphisms) between them which can be empty

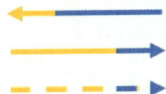

Each arrow has source (domain) and target (codomain) objects

Arrows are subject to the condition of compositionality

$X, Y, Z \in Ob(C)$
$f, g \in Ar(C)$
$f \in Ar(X, Y)$
$g \in Ar(Y, Z)$

$$X \xrightarrow{f} Y \xrightarrow{g} Z$$

$$X \xrightarrow{g \circ f} Z$$

$$X \xrightarrow{f} Y \xrightarrow{g} Z$$

$$X \xrightarrow{g \circ f} Z$$

The composition of arrows is associative: $f \circ (g \circ h) = (f \circ g) \circ h$

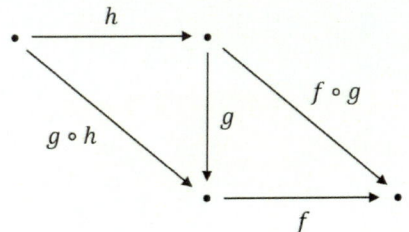
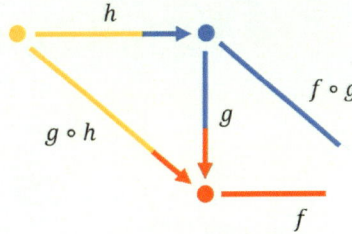

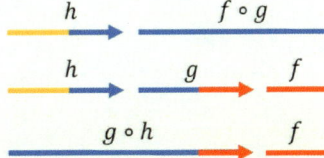

For each $X \in Ob(C)$ there is a special arrow $id_X \in Ar(X,X)$ that satisfies the following properties: $f \circ id_X = f$ and $id_X \circ g = g$

$$X \xrightarrow{id_X} X \xrightarrow{f} Y$$

$$Y \xrightarrow{f^{-1}} X \xrightarrow{id_X} X$$

$$id_X \quad X$$

$$g = f^{-1}$$

$$id_X \quad X$$

$$X \xrightarrow{id_X} X \xrightarrow{f} Y$$

$$Y \xrightarrow{f^{-1}} X \xrightarrow{id_X} X$$

$R \in Ob(\mathbf{C})$ is a retract of $X \in Ob(\mathbf{C})$
$s, r \in Ar(\mathbf{C})$ are section and retraction

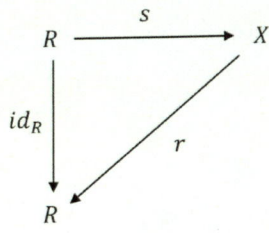

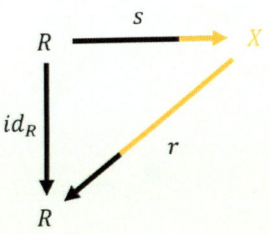

$Ar(C) \ni f: X \to Y$ is an equivalence if there is
$Ar(C) \ni g: Y \to X$ with $g \circ f = id_X$ and $f \circ g = id_Y$

$$X \xrightarrow{f} Y \xrightarrow{g} X$$

$$Y \xrightarrow{g} X \xrightarrow{f} Y$$

$$X \xrightarrow{f} Y \xrightarrow{g} X$$

$$Y \xrightarrow{g} X \xrightarrow{f} Y$$

A covariant functor between categories $F: \mathbf{C} \to \mathbf{D}$
$F: Ob(\mathbf{C}) \to Ob(\mathbf{D})$ and
$F: Ar(\mathbf{C}) \to Ar(\mathbf{D})$ with $F: Ar(X,Y) \to Ar(F(X), F(Y))$

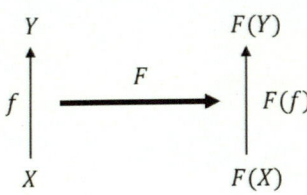

A contravariant functor between categories $G: \mathbf{C} \to \mathbf{D}$
$G: Ob(\mathbf{C}) \to Ob(\mathbf{D})$ and
$G: Ar(\mathbf{C}) \to Ar(\mathbf{D})$ with $G: Ar(X,Y) \to Ar(G(Y), G(X))$

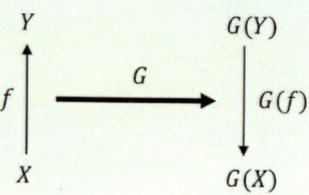

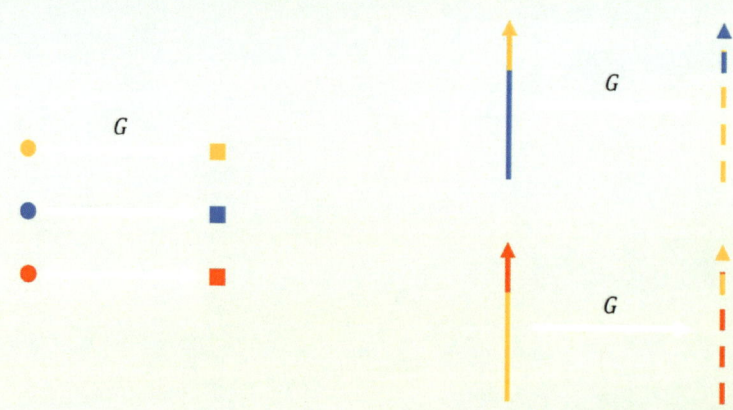

A covariant functor F and contravariant functor G

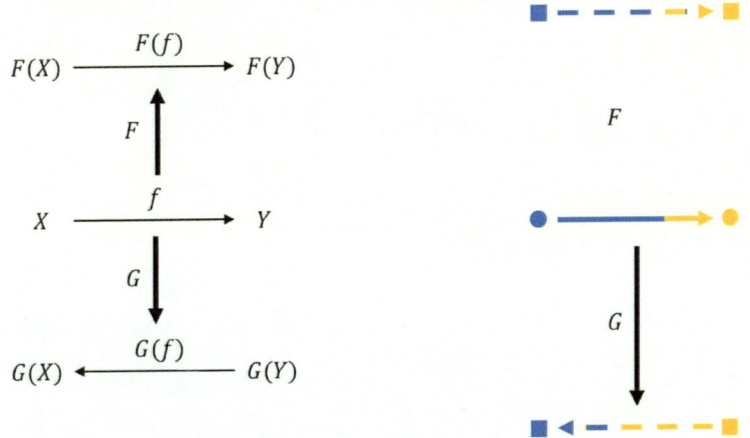

A covariant functor F and contravariant functor G
$F(g \circ f) = F(g) \circ F(f)$ and $G(g \circ f) = G(f) \circ G(g)$

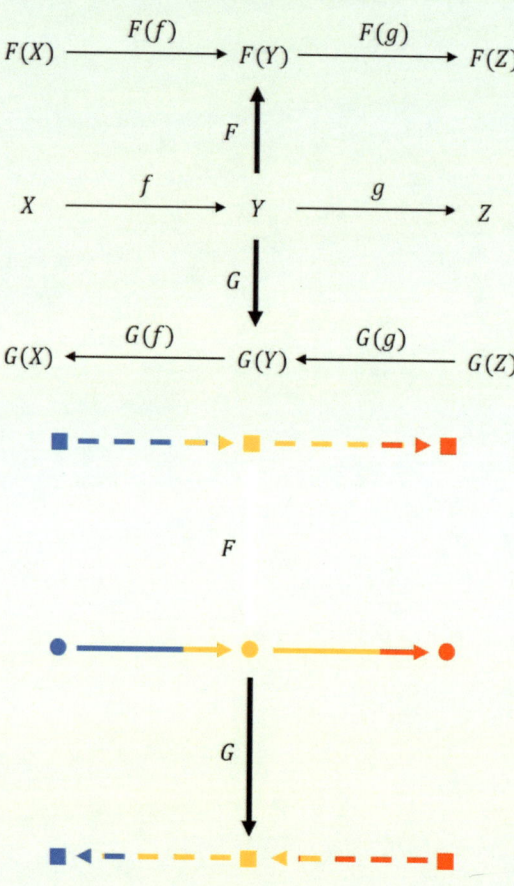

A natural transformation N is a collection of arrows between functors' targets indexed by functors' source objects
$n_X \in Ar(F(X), G(X))$

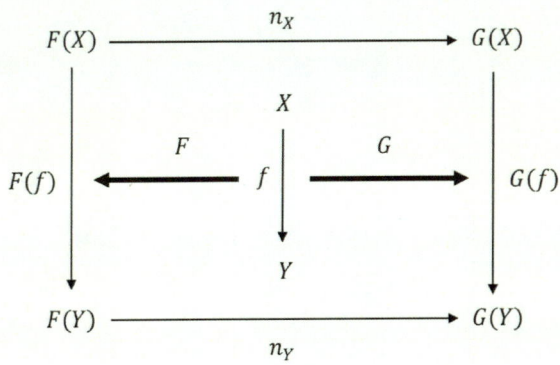

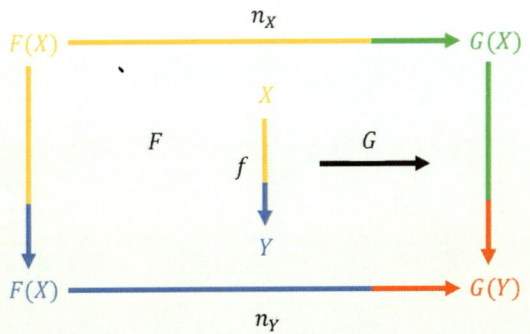

A 2-category is a category with arrows having a category structure: arrows as objects (1-cells) and arrows (2-cells) between 1-cells

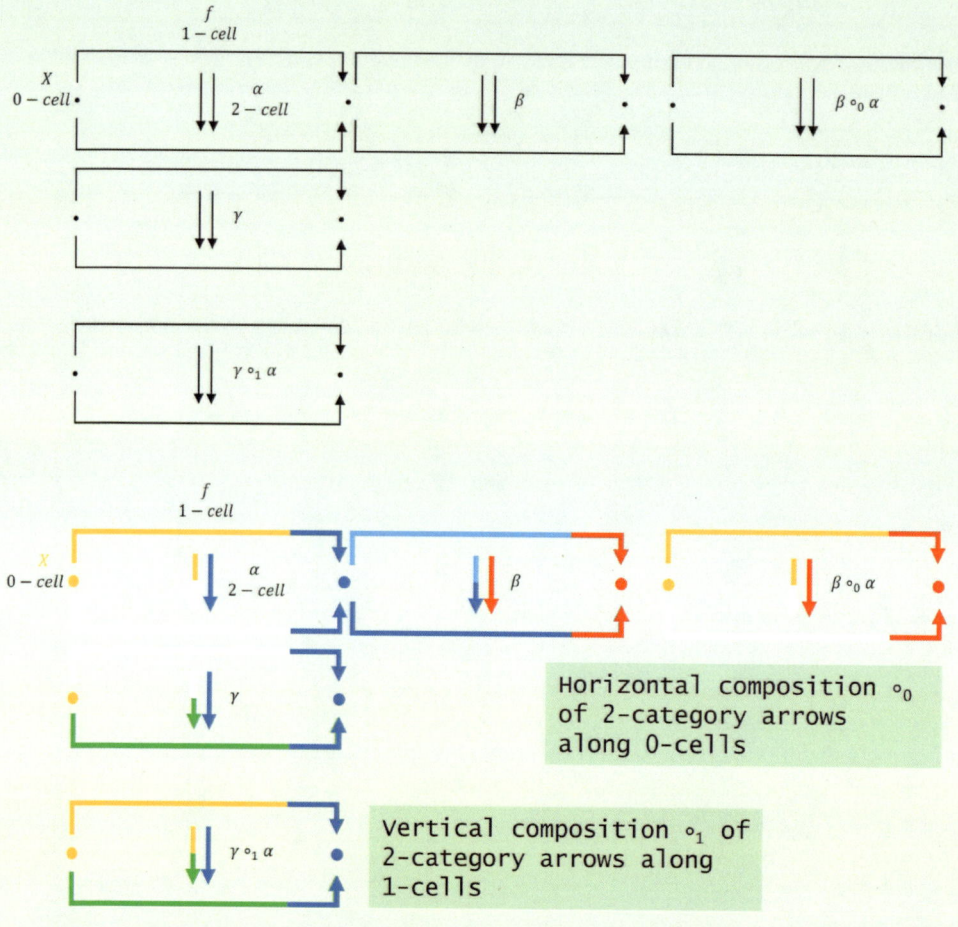

Horizontal composition $\circ_0$ of 2-category arrows along 0-cells

Vertical composition $\circ_1$ of 2-category arrows along 1-cells

www.ingramcontent.com/pod-product-compliance
Lightning Source LLC
Chambersburg PA
CBRC101955190326
41520CB00006B/227